Instruments of a Delicate Character

The Annotated Lecture from 1912

E. W. Hobson

Response Essay by J. Brian Pitts
Edited by Tristan Abbey

Science Venerable Press
Tampa, Florida
2024

Cover art: Sandro Botticelli
“Saint Augustin dans son cabinet de travail” (ca. 1480)
Wikimedia/Pubic Domain

ISBN: 978-1-959976-06-6

Published by Science Venerable Press, an imprint of Comarus Analytics LLC,
Tampa, Florida

www.sciencevenerable.com

Interior designed by Amnet
Cover design by Derek Thornton / Notch Design
Series design by Nathaniel Roy

"I have said that Mathematics is the oldest of the sciences; a glance at its more recent history will show that it has the energy of perpetual youth."

— Ernest William Hobson, 1910[1]

1 E. W. Hobson, Presidential Address to Section A, *Report of the Eightieth Meeting* (Sheffield), British Association for the Advancement of Science (John Murray, 1911), pp. 509-522; p. 511.

Table of Contents

Preface

Ernest William Hobson (1856-1933) did all the right things to distinguish himself as a British mathematician. He joined—or, rather, was "elected"—to the Royal Society in 1893 and later served as president of the London Mathematical Society. A true man of Cambridge, where he studied, taught, and researched, Hobson went on to earn prestigious medals in mathematics and published several books in philosophy, mathematical history, and theory. Hobson was a contemporary of Andrew Russell Forsyth (see *The Silent Science*, published by the Science Venerable Press in 2024), and in fact took over Forsyth's role as Sadleirian Professor of Pure Mathematics in 1910. He was himself succeeded by the far more famous G. H. Hardy in 1931, who incidentally wrote his obituary for *Nature* and the Royal Society.[2]

Instruments of a Delicate Character is a simple booklet. It contains a newly annotated version of Hobson's lecture, "Mathematics, from the points of view of the Mathematician and of the Physicist," delivered to University College London in 1912. The cumbersome title of the lecture has been replaced here by an excerpt from it. I am also

2 See G. H. Hardy, "Ernest William Hobson (1856-1933)," *Biographical Memoirs*, Vol. 1, Issue 3 (Royal Society, December 1934), pp. 236-249.

delighted to feature a response essay by J. Brian Pitts, philosopher-physicist at Cambridge.

As with nearly all previous Science Venerable Press publications, the MacTutor History of Mathematics Archive at the University of St Andrews proved an enduring, wonderful resource. HathiTrust Digital Library, Perseus Digital Library, Wikisource, and the Internet Archive also remain invaluable aids.

Tristan Abbey
Tampa, Florida

Ernest William Hobson
(Credit: Royal Society)

Mathematics, from the points of view of the Mathematician and of the Physicist [1912]

An address delivered to the Mathematical and Physical Society of University College, London

by

E. W. HOBSON, Sc.D., LL.D., F.R.S.[1]

Sadleirian Professor of Pure Mathematics in the University of Cambridge

1 Doctor of Science; Legum Doctor; Fellow of the Royal Society.

I

IF WE WERE to question a man of average education, or even one of the majority of those who belong to the cultured class, as to what he conceives to be the nature of Mathematical Science, and as to what he thinks are the aims of those who cultivate it, we should probably receive a somewhat vague answer, conveying his impression that Mathematics is concerned with calculations carried on by means of processes involving a copious use of symbols and diagrams entirely unintelligible to the uninitiated. He might not improbably add that such symbols and diagrams are of no interest to anyone except a few individuals who have an unaccountable taste for such things, and happen to be endowed with a peculiar transcendental faculty the possession of which is quite unnecessary for other people, and which he himself does not possess, and moreover that he is quite happy without it.

If pressed, our friend would probably admit that there are subjects, such as Arithmetic and Mensuration,[2] which are extremely useful, and some knowledge of which is essential for various purposes, but that such subjects are hardly Mathematics, and that anything beyond them is of little or no concern to the world in general.

---

2 "Mensuration" is applied geometry, the science of measurement.

If, as is likely, he had himself, at school, studied a little Geometry and Algebra, he would say that the former is largely concerned with difficult and tedious proofs of things that are either obvious or meaningless, and that the latter is concerned with juggling with letters in an extremely troublesome manner, and for no ascertainable object.

Even if we obtained a somewhat less crude answer to our question, that answer would probably be, in a greater or less degree, an inadequate one as a characterization of the essential features of Mathematical Science. In the scientific world in general, and even among students of those branches of Science in which Mathematical thought has played an exceptionally large part, decidedly vague and narrow conceptions of the functions of Mathematical thinking are current. In such circles, the notion is extremely common that the sole function of Mathematics is to provide the means of carrying out whatever calculations may be necessary for the special purposes of each department of Physical Science, theoretical or applied, and thus that Mathematics plays in them a comparatively humble part analogous to that of a mechanical tool.

As a matter of fact Mathematical thinking has done very much more than merely to provide methods of calculating; it has played a most important part in the formation of the concepts with which the Physical Sciences work; it has decided in many cases for them not only how to calculate but what the things are that they ought to calculate; it has reduced the originally vague conceptions which arise in connection with physical observation to precise forms in which they can be exhibited as measurable quantities.

Mathematical thinking, in a more or less explicit form, pervades every department of human activity. The grocer, when he weighs out his sugar, makes use of Mathematical conceptions which were developed only by a long process of evolution; when he enters his receipts in his books, the notation he uses is a warrant of the importance of Mathematical form, and embodies one of the triumphs of our race as a mode of economizing thought. The Engineer, when he makes the necessary graphical calculations of the stresses in the bridge he is building, has to employ concepts and methods belonging to a decidedly advanced state of Mathematical thought. The Philosopher, in his reflections on spatial and temporal relations, on number and quantity, on matter and motion, is in a region of thought in which the boundary between his own domain and that of the Mathematician is almost non-existent. The Epistemologist has always to take Mathematical knowledge as a kind of touchstone on which to test his theories of the nature of knowledge. The dominant views in various departments of philosophical thinking have been modified in important points by the results of recent Mathematical research, and will, I think, in the future, not improbably be further modified from the same quarter.

Mathematical thought is at once the most all-pervading and the most highly specialized department of mental activity.

The origin of Mathematical thinking, and the reason of its ubiquity in all departments of Science, and even of practical life, are not far to seek. The physical world in which we live appears to us as a manifold of objects extended in space,[3] their spatial relations

3 "Space is a manifold, but not every manifold is a space." Paul Carus, *The Foundations of Mathematics: A Contribution to the Philosophy of Geometry* (Open Court Publishing, 1908), p. 108. The four classical manifolds are a line (one dimension), plane (two dimensions), space (three dimensions), and space-time (four dimensions).

exhibiting at any one time an endless variety, and varying at different times. The primary Mathematical operations of counting and measuring, leading up to the concepts of number and geometric form, took their first beginnings in the practical interests of man living in that world. These Mathematical concepts, and the methods of dealing with them, underwent a gradual evolution in close connection with the practical efforts required to grasp, classify, and characterize, and so far as possible to dominate, the spatial relations of actual bodies, both in their fixity and as changing in temporal succession. The more closely men scrutinized natural phenomena, at first for practical reasons, and later from intellectual curiosity, the more things and processes they found to have aspects which are measurable, and the more they were able to employ their developing Mathematical processes and concepts for the precise characterization of various aspects of the world of phenomena. But, as I hope to be able to show in some detail, the natural development of Mathematical thought, starting as it did in connection with the more obvious aspects of sensuous experience, under the pressure of physical needs, brings it to a region reaching far beyond that in which the primitive intuitions of time, space, and matter formed the exclusive subject matter of the Science.

Mathematics is not merely a subject to be studied and developed by a class of specialists; it has to play a part, and usually an important and indispensable part, in other departments of Science; and consequently there are various attitudes taken up with regard to Mathematics, dependent on the relations of the subject to these various departments.

I have chosen for my subject this evening a comparison of the ways in which the nature and functions of Mathematics are regarded, and should be regarded, by the Mathematician proper, and by the Physicist. I deal with the subject primarily as a comparison of the attitudes of those who have been engaged in the past, and of those who are now engaged, in furthering the progress of Mathematics and Physics as growing branches of knowledge: but if it be true, as I think it is, that the spirit in which a student of any branch of Science pursues his work should approximate as much as possible to that spirit which animates the researcher, the kind of comparison which I here attempt may, I hope, be not without interest or utility to the younger students of either branch of knowledge.

II

For many purposes, the Astronomer, the Engineer, the Statistician, and the Biometrist[4] might be grouped with the Physicist, as having analogous, although by no means identical, dealings with the Mathematician. For example, the Engineer, like the Physicist, has constantly to make use of Mathematical methods; but as his ultimate aim is to harness the forces of nature and use them to obtain practical results, rather than to bring their relations under general laws and concepts as the theoretical Physicist does, he is perhaps less directly concerned than the Physicist with the part which Mathematical Science has played in the formation of the concepts with which the theoretical Physicist works. He uses applied Mathematics, and applied Physics, and is apt to take both of them more or less as ready-made products, although he cannot do so beyond a certain point without grave danger to his efficiency as a scientific engineer.

In former times the Mathematician and the Physicist were usually one and the same man.[5] Even as late as the eighteenth century

4 Biometrics today are typically understood as personal identifiers in a security context. Here, the term "biometrist" equates to an early instance of biostatistics and bioinformatics.

5 Archimedes, Ibn al-Haytham, Pascal, Huygens, Newton, Leibniz.

this was very generally the case; it was in the nineteenth century that the increasing complexity of both Sciences produced that separation of the two departments which has become continually more marked, and has reached its extreme point in our own time. To attempt to give a complete explanation of this phenomenon would be to attempt to write the history of both Sciences. I can only give, in the course of my remarks, some brief indications of the causes, lying in the lines of development of both Sciences, which have brought about this separation.

The typical Mathematician and the typical Physicist of to-day have their centres of interest very differently located. This difference, though inevitable in view of the exigencies of the two studies, and to a large extent advantageous, has of course serious drawbacks. The chief drawback is that each specialist, from lack of interest in, and knowledge of, the progress of the other great department, is apt to miss that large source of inspiration in his own study which is supplied by the other one.

The essential difference of the ways in which Mathematics is contemplated by the Pure Mathematician and by the Physicist sometimes comes strikingly into expression. I remember, some quarter of a century ago, at a Board meeting at Cambridge, the subject of [Friedrich] Bessel's functions came into the discussion, I think in connection with some proposal to include them in an examination syllabus. Their utility in connection with Applied Mathematics having been referred to, a very great Pure Mathematician[6] who was present ejaculated—"Yes, Bessel's functions are very beautiful functions, in spite of their having practical applications." It would

6 This unnamed mathematician could have been Arthur Cayley.

have been interesting to have heard what this great man would have said if he had known that Professor [John] Perry would one day propose the desecration of these beautiful functions by recommending them as suitable playthings for young boys.[7]

The typical Pure Mathematician is quite satisfied with a result which is expressed with the requisite precision by the employment of an apparatus of symbols each of which has been properly defined. To the typical Physicist, on the other hand, a formula seems often to have very little meaning unless he can interpret it in some intuitional manner that gives it a meaning which in its purely abstract form it does not possess for him.

This divergence of mental habits appears in a striking way in some remarks made by that ultra-Pure Mathematician [James Joseph] Sylvester about a treatment of the subject of Spherical Harmonics by the theory of poles, that was given by [James Clerk] Maxwell, a man with physical instincts of the most pronounced order. Maxwell wrote of his own theory:

> In numerical investigations I have often been perplexed on account of the apparent want of definiteness of the idea of a

---

7 Reference to an address by John Perry on September 14, 1901, at the Glasgow meeting of the British Association. Perry was a Fellow of the Royal Society and on faculty at the Royal College of Science. "I should have thought that the advantage of knowing how to use spherical harmonics or Bessel functions at the age of seventeen, so as to be able to start in mathematics at Cambridge just about the place where some of the best mathematical men now end their studies for ever, of starting at this high level with youthful enthusiasm, and individuality, and inventiveness, would more than compensate for the evils of skipping [more elementary material]." Reprinted in *Discussion on the Teaching of Mathematics*, British Association (Macmillan and Co., 1902), p. 13.

> [Pierre-Simon] Laplace's Coefficient or spherical harmonic. By conceiving it as derived by the successive differentiation of $\frac{1}{r}$ with respect to i■ axes, and as expressed in terms of the positions of its i■ poles on a sphere, I have made the conception of the general spherical harmonic of any integral degree perfectly definite to myself, and I hope also to those who may have felt the vagueness of some other forms of the expression.[8]

Commenting upon this, Sylvester writes:

> The method of poles for representing spherico-harmonics, devised or developed by Professor Maxwell, really amounts to neither more nor less than the choice of an apt canonical form for a ternary quantic, subject to the condition that the sum of the squares of its variables (here differential operators) is zero; and I am quite at a loss to understand how it can at all assist 'in making the conception of the general spherical harmonic of an integral degree perfectly definite,' or what want of definiteness apart from the use of this canonical form can be said to exist in the subject.[9]

---

8 James Clerk Maxwell, *A Treatise on Electricity and Magnetism*, Vol. I (Clarendon Press, 1873), pp. 162-163.

9 J. J. Sylvester, "Note on Spherical Harmonics," *Philosophical Magazine and Journal of Science*, Vol. II, Series 5, No. 11 (Taylor and Francis, October 1876), pp. 291-307; p. 305.

But the spirit of the great Algebraist had been further grieved by Maxwell in this matter, for he writes:

> With all possible respect for Professor Maxwell's great ability, I must own that to deduce purely analytical properties of spherical harmonics, as he has done, from "Green's theorem" and the "principle of potential energy" [Electricity and Magnetism, vol. I., p. 168], seems to me a proceeding at variance with sound method, and of the same kind and as reasonable as if one should set about to deduce the binomial theorem from the law of virtual velocities, or make the rule for the extraction of the square root flow as a consequence from Archimedes' law of floating bodies.[10]

Speaking to some extent from personal experience, one of the effects of prolonged study of some of the more abstract branches of Mathematics, as for example the Theory of Functions, is that one begins to take the greatest interest in, and to be most attracted by, just those aspects of the subject which are most remote from the interests of the Physicist. One gets into an attitude of mind in which the kind of well-behaved functions, without abnormal singularities, that are likely to turn up in Physical investigations, appear to have a somewhat bourgeois aspect, in their comparatively uninteresting respectability.

In the mind of one who makes a minute and prolonged study of the peculiarities which [Joseph] Fourier's series may present, a not improbable effect of that study is that the ordinary Fourier's

10 Ibid., p. 306.

series, which converge everywhere quite normally, begin to acquire a certain tameness of aspect which deprives them of interest. The failure of convergence of [Brook] Taylor's series becomes to some Mathematical students a matter of greater interest than that presented by the series in the ordinary cases in which, by their regular convergence, they are fitted for purposes of application.

III

I HAVE ALREADY called attention to the fact that Mathematical thinking has played a very important part in the formation of the fundamental concepts of the Physicist; very often this part has been a dominant one. Many of these concepts could only have received a precise meaning, and could only have taken definite forms, as the result of the work of Mathematicians, and their formulation often appeared as the result of a long train of previous Mathematical thinking.

For example, the conception of Energy, and the exact meaning of the great generalization known to us as the law of the Conservation of Energy, emerged as results of the development of the abstract side of molar mechanics,[11] which determined the mode in which the kinetic energy of moving bodies and potential energy as work are defined as measurable quantities. Only by the transference and extension of these notions to the molecular domain did the

11 In chemistry, $6.02214076 \times 10^{23}$ molecules = 1 mole. "By a molar body we mean one which is revealed to the senses without the use of instruments such as the telescope, the microscope, the spectroscope, or the crucible, aided by computation and logical ideation." John Wesley Powell, *Truth and Error, or the Science of Intellection* (Open Court Publishing Company, 1898), p. 129.

conception involved in the modern doctrine become possible. The doctrine of the conservation of energy for the case of molar bodies had been established before [James Prescott] Joule and [Julius von] Mayer commenced their work, and was a necessary presupposition of their further development. Joule was able to determine the mechanical equivalent of heat[12] only owing to the fact that mechanical work was already regarded as a measurable quantity, measured in a manner which had been fixed in the course of the development of the older Mathematical Mechanics.[13]

The notion of Potential, fundamental in Electrical Science, and which every Physicist, and every Electrical Engineer, constantly employs, was first developed as a Mathematical conception during the eighteenth century in connection with the theory of the attractions of gravitating bodies.[14] It was transferred to the electrical domain by George Green[15] and others, together with a good deal of detailed mathematics connected with it which had previously been applied to the gravitational potential function.

12 "Indeed the phenomena of nature, whether mechanical, chemical, or vital, consist almost entirely in a continual conversion of attraction through space, living force [*vis viva*], and heat into one another." James Prescott Joule, "On Matter, Living Force, and Heat," *Scientific Papers*, Vol. I (Physical Society of London, 1884), pp. 265-276; p. 273.

13 These older mechanics were developed by Descartes, Leibniz, Coriolis, Poncelet, and Carnot.

14 Early work on potential function was done by Newton, Laplace, and Lagrange.

15 George Green, *An Essay on the Application of Mathematical Analysis to the Theories of Electricity and Magnetism* [1828], in *Mathematical Papers of the Late George Green*, edited by N. M. Ferrers (Macmillan and Co., 1871), pp. 1-82.

The ultimate aim of the Physicist, even of one whose work is almost exclusively experimental, is much higher than that of attaining to a merely empirical knowledge of facts. His real object is to classify facts in such a way as to refer them to general laws which are of a more or less abstract character, and which involve concepts of schematic representations that require, for the fixing of their exact denotation, the aid of the Mathematician. The man of true Physical instincts, endowed with the great faculty of scientific imagination, possessed for example by Lord Kelvin in a very remarkable degree, is forever imagining models which shall enable him by their working to represent and depict the course of actual physical processes. The possibility and consistency of such models require Mathematical Analysis for their investigation. The Mathematician may also, by tracing the necessary consequences of the postulation of a model of a particular type, formulate crucial tests in accordance with which further experiments will decide whether a particular type of model can be retained at least provisionally, or whether it must be rejected as inadequate for the representation of known facts, and must give place to some other model of a different type.

Perhaps the most striking example of the services which have been rendered to Science by the contemplation of various models, many or all of which have ultimately been found to be inadequate for complete representation, is to be found in the history of Optics. The various forms of the corpuscular theory, and of the wave theory of Light were all attempts to represent the phenomena by models, the value of which had to be estimated by developing their Mathematical consequences, and comparing these consequences with the results of experiments. The adynamical[16] theory of [Augustin-Jean] Fresnel, the elastic solid theory of the ether developed by [Claude-Louis] Navier, [Augustin-Louis] Cauchy, [Siméon Denis] Poisson, and Green, the labile ether theory developed by Cauchy and Kelvin, and the rotational ether theory of [James] MacCullagh were all efforts of the kind I have indicated; they were all successful in some greater or less degree in the representation of the phenomena, and they all stimulated Physicists to further efforts to obtain more minute knowledge of those phenomena.[17] Even such an inadequate theory as that of Fresnel led to the very interesting observation by Humphrey Lloyd of the phenomenon of conical refraction in crystals, as the result of the prediction by [William] Rowan Hamilton that the phenomenon was a necessary consequence of the Mathematical fact that Fresnel's wave surface in a biaxial crystal possesses four conical points.[18]

---

16 "Fresnel's investigation can scarcely be called a dynamical theory in the strict sense, as the qualities of the medium are not defined. His method was to work backwards from the known properties of light, in the hope of arriving at a mechanism to which they could be attributed; he succeeded in accounting for the phenomena in terms of a few simple principles, but was not able to specify an aether which would in turn account for these principles." E. T. Whittaker, *A History of the Theories of Aether and Electricity*, Vol. I (Longmans, Green, and Co., 1910), p. 135.

17 Ibid. Whittaker recounts these tales in Chapter IV and V. The only thing everybody could agree on—at least, until Einstein's foray in 1905—was that the ether existed.

18 Fresnel, Augustin. *Mémoire sur la Double Refraction*, in *Mémoires de L'Académie Royale des Sciences* Tome VII (Chez Firmin Didot, 1827), pp. 45-176. Hamilton, William Rowan. "On Some Results of the View of a

Although the theoretical Physicist has for his real aim the formulation of abstract schemes for the description and correlation of the physical phenomena which he observes, with the Mathematician processes of abstraction must go very much further than with the Physicist. A strong tendency of Mathematics in its later developments is to split up notions, originally undivided, into components, and to proceed to deal with these components in isolation, and often in separate branches of study. For example, during the last half-century, number and measurable quantity have been separated from one another; the idea of number alone has been recognized as the foundation upon which Mathematical Analysis rests, and the theory of extensive magnitude is now regarded as a separate department in which the methods of Analysis are applicable, but as no longer forming part of the foundation upon which Analysis itself rests. For the purposes of analysing the implications of the methods employed, and of pushing those methods to the highest possible degree of development, this kind of separation is indispensable, and has led to the very abstract form in which much modern Mathematics is exhibited.

For the Physicist, on the other hand, it is essential that abstraction should not go nearly so far as the Mathematician must push it. Too much abstraction on the part of the Physicist would entail the penalty that he would lose his way in a field which is barren for his

---

Characteristic Function in Optics," *Report of the Third Meeting of the British Association for the Advancement of Science* (John Murray and Sons, 1834), pp. 360-370. For the experimental confirmation, see Humphrey Lloyd, "On Conical Refraction," pp. 370-373, and Lloyd, "On the Phenomena Presented by Light in Its Passage along the Axes of Biaxial Crystals," *Transactions of the Royal Irish Academy*, Vol. XVII (1831 [1833]), pp. 145-157.

purposes, and would lead to a loss of contact with the phenomenal world. To do what the Mathematician does, and must do, in view of the essential interests of his study, would be fatal to the Physicist, to whom above all things a large degree of concreteness in his conceptions is indispensable.

IV

NOT ONLY DID Mathematical Science take its origin in the necessities and interests of man living in the physical world, but at every stage of its development the problems of Physics have been the source of the ideas which have directed the Mathematician, and from which new paths of investigation have been suggested to him. But every great problem which has been placed before the Mathematician from the physical side of things has given rise to a train of ideas which the law of his being compelled him to pursue, and has started a host of questions to which he was impelled to find the answer. These have led him in most cases far beyond the original domain in which the problems originated.

The most abstract branches of modern Mathematics, the theory of functions, real and complex, the theories of groups and of differential equations—all arose originally from physical beginnings, but have reached out into vast developments, apparently, but not by any means always really, remote from the physical region. At any moment one or other of these developments may become urgently necessary for the purposes of Physics, and may thus be in a position to pay back some of the debt they owe to the parent from whose side they have wandered so far.

The question is often asked, in some form or other, why Mathematicians cannot restrict themselves more to those aspects of their Science which bring them in contact with Physics, and which are concerned with what often receives the question-begging and ambiguous name of *reality*; a word that has an indefinite number of shades of meaning, varying with every difference in philosophical view, but which in this connection is generally associated with that kind of reality supposed exclusively to appertain to the physical world.

Why, it is asked, do modern Mathematicians to so great an extent as at present wander away from the source in which their Science had its origin, and from which its ever-renewed inspiration has been received, in order to lose themselves in a transcendentalism which, in its aloofness from physical investigation, condemns them to an endless and barren immersion in abstractions of their own creation? Why has modern Mathematics cut itself off from the roots of the Science?

Well, the answer to this often-recurring question is that the modern abstract lines of thought of Mathematicians lie in the natural line of progress for the minds of those who are unable to rest content when questions force themselves upon their attention, without making the most strenuous endeavours to find answers to those questions. To stop short at a point dictated by considerations of applicability to Physics is impossible to those to whom clear and thoroughly defined conceptions are a desideratum, the lack of which in any department of their study leaves them no rest. To attempt to confine the activities of Mathematicians by imposing upon them a restriction of the nature indicated in the above question would be to attempt to strangle the Science as a progressive development. Mathematics can in the long run be developed to the highest degree of perfection, not only from the point of view of specialists within its own domain, but also as constituting an essential component of the intellectual life and stock of ideas of the world, only on the condition that it is allowed full freedom of self-expression. The utilitarian notion, in this connection as in so many others, has the fatal limitation that it attempts to assign limits to what is, or may in the future become, useful, in accordance with a more or less arbitrarily restricted standard of what constitutes utility.

When the exigencies of Physics suggest to the Mathematician some process, or some special problem for solution, he is impelled to search for some generalization, some law, under which a whole class of analogous processes or problems can be subsumed. He is never content with having obtained merely isolated results; general methods and general laws of which the isolated results are but special cases are what he invariably looks for.

The Physicist also pursues the same course; he, too, is really occupied in attempting to exhibit isolated phenomena as particular cases of some general law under which a whole class of phenomena can be subsumed. A narrow utilitarianism would be as fatal to the growth of Physics as to that of Mathematics. The Physicists of the eighteenth century, when they examined the phenomena of Electricity and of Magnetism, before any relation between these two sets of phenomena had been discovered, were actuated solely by intellectual curiosity, which impelled them to attempt the discovery of laws for the correlation of the isolated facts known to them. Had their spirit been utilitarian in the narrow sense, they would probably have left the matter on one side, as holding out no prospect of useful application; and one result of such abstention would have been that we should not now possess the electric telegraph, the telephone, or the motor-car.[19]

The most abstract Mathematical investigators, the most purely experimental Physicist, and every investigator whose path lies between the two extremes—all are taking humble shares in the great enterprise of bringing, in whatever measure may prove possible, processes not only of what we call the physical world, but also of what we call the psychical world, under the categories of conceptual thought. It would be rash indeed to attempt to estimate with any accuracy the ultimate values of the shares contributed by different kinds of investigators, with all their varied shades of mental leanings, towards this great work. Mathematicians of all shades of interest, and Physicists of all kinds as regards their attitude towards the more abstract and the more concrete aspects of their subject, are required as contributors to the supreme work, the attainment of the underlying purpose of which demands the sustained efforts of minds of the most varied types through the ages of the history of the human race.

19 Priority disputes abound, but classically Samuel Morse transmitted his first message via telegraph in 1844, Alexander Graham Bell placed his first phone call in 1876, and Karl Benz received his patent for the Motorcar in 1886.

V

The Mathematical Physicist plays a part of supreme importance as an intermediary and interpreter between the Pure Mathematician and the experimental Physicist. In order that he may do his work effectually he must follow with an alert eye the lines of progress both in Mathematics proper, and in experimental Physics. The nature and the amount of the work he can do vary much with the character of the current physical investigations.

It is probably true that, in spite of some brilliant exceptions, the Mathematical Physicist does not just at present take as prominent a part as was formerly the case, especially during the nineteenth century, the age of [James Clerk] Maxwell, Kelvin, [George Gabriel] Stokes, [and Hermann von] Helmholtz. In attempting to assign reasons which may explain this partial eclipse of the activities of the Mathematical Physicist, it appears possible that a part of the explanation may be afforded by the actual separation in education of the training of Physicists and Mathematicians. Although this separation is probably to some extent necessary in the highly specialized state of both departments which now subsists, and as the differing tastes of students determine whether their studies shall be in the main experimental or in the main Mathematical, the separation has been in practice carried much too far. No doubt more might be done than at present by our educational institutions to coordinate Mathematical and Physical studies so as to train up a race of Mathematical Physicists. I am inclined, however, to think that the main ground of the present comparatively smaller part played in scientific life by the Mathematical Physicist is to be found in the present trend of Physical investigations, and that probably the present state of things is only a temporary one.

In earlier times, when physical investigations were principally concerned with the grosser and more obvious phenomena, when molar Mechanics, and especially celestial Mechanics, occupied the centre of the interests of physical investigation, the passage from the observation of concrete phenomena to their abstract Mathematical representation was comparatively easy. The observational work was simpler and less technical than at present; highly equipped physical laboratories had not yet come into existence and were not then necessary. The Mathematical Physicists and Astronomers of the eighteenth century were largely engaged in working out the detailed implications of the law of gravitation, and had commenced, largely under the influence of the idea of action at a distance, to work out problems such as those connected with the vibrations of strings and other bodies.

In the earlier part of the nineteenth century the centre of physical interest passed on to such subjects as Hydrodynamics, the Conduction of Heat, and Elasticity, in which an abstract representation of a body as a continuous plenum was sufficient for many purposes, and made the problems readily accessible to continuous Mathematical Analysis. During this period much attention was given to the more obvious phenomena of Electricity and Magnetism, and much of the Mathematical Analysis which had been devised for the purpose of dealing with problems of gravitational attractions, vibrations, etc., was found, with further development, to be applicable to the new problems which arose. Much of the work, such as that of [André-Marie] Ampère in connection with the ponderomotive forces due to electric circuits,[20] was still carried out under the influence of the idea of action at a distance, first brought into prominence in connection with the Newtonian law of gravitation, but the idea of the continuous medium gradually became the dominating notion.

The period in which Physical Mathematics was applied with such great success to continuous media probably reached its culmination in Maxwell's equations of Electrodynamics which are now usually regarded as representing the average effects exhibited when actual discreteness is smoothed out. In our own time the centre of physical interest has transferred itself, in connection with Electromagnetism, to the molecular and sub-molecular domain, in which discrete objects become the subject of scrutiny. The boundaries between Physics and Chemistry have been broken down. In this region of investigation the perseverance, skill, and ingenuity of Physicists working in well-equipped laboratories have been rewarded by the discovery, during the last two decades, of a crowd of remarkable facts, probably destined to have the most far-reaching influence upon our conceptions of the material world. The sifting and interpretation of these facts is at present most incomplete, and will in the future give rise to many problems in grappling with which work in abundance will be provided for the Mathematician.

---

20 André-Marie Ampère, *Recueil D'Observations Électro-Dynamiques* (Chez Crochard, 1822).

As yet the statement of these problems is in many cases not sufficiently precise and definite to allow of translation into terms with which the Mathematician can readily deal. It seems however highly probable that, when Mathematics gets a grip of these problems, questions will arise which will tax to the utmost the resources of Mathematical Science.

VI

It would be rash to attempt to limit the kinds of Mathematical processes which may in good time be made available for application to the new and difficult problems which can be discerned at present as likely to arise in the not very distant future. It has been said that the Theory of Numbers is a subject which has never been soiled by any practical application. Who can be absolutely sure that even so apparently transcendental a branch of thought as this will always remain undefiled by the contaminating touch of physical application?

In the history of Science it is possible to find many cases in which the tendency of Mathematics to express itself in the most abstract forms has proved to be of ultimate service in the physical order of ideas. Perhaps the most striking example is to be found in the development of abstract Dynamics. The greatest treatise which the world has seen, on this subject, is [Joseph-Louis] Lagrange's *Mécanique Analytique*, published in 1788. In order to characterize the spirit in which this great work was conceived I cannot do better than quote the words of Lagrange himself from the Preface. He writes:

> We have already various treatises on Mechanics, but the plan of this one is entirely new. I intend to reduce the theory

> of this Science, and the art of solving problems relating to it, to general formulae, the simple development of which provides all the equations necessary for the solution of each problem. I hope that the manner in which I have tried to attain this object will leave nothing to be desired...
>
> No diagrams will be found in this work. The methods that I explain require neither geometrical, nor mechanical, constructions or reasoning, but only algebraical operations in accordance with regular and uniform procedure. Those who love Analysis will see with pleasure that Mechanics has become a branch of it, and will be grateful to me for having thus extended its domain.[21]

If ever a work was conceived in the purely abstract Mathematical spirit, it is surely this one. Well, let us see what came of it.

Lagrange's idea of reducing the investigation of the motion of a dynamical system to a form dependent upon a single function of

---

21 "On a déjà plusieurs Traités de Méchanique, mais le plan de celui-ci est entièrement neuf. Je me suis proposé de réduire la théorie de cette Science, et l'art de résoudre les problèmes qui s'y rapportent, a des formules générales, dont le simple développement donne toutes les équations nécessaires pour la solution de chaque problème. J'espère que la manière dont j'ai tâché de remplir cet objet ne laissera rien à désirer...
"On ne trouvera point de Figures dans cet Ouvrage. Les méthodes que j'y expose ne demandent ni constructions, ni raisonnements géométriques ou méchaniques, mais seulement des opérations algébriques, assujetties à une marche régulière et uniforme. Ceux qui aiment l'Analyse verront avec plaisir la Méchanique en devenir une nouvelle branche, et me sauront gré d'en avoir étendu ainsi le domaine."
See *Oeuvres de Lagrange*, Tome Onzième (Gauthier-Villars et Fils, 1888), pp. xi-xii.

the generalized coordinates of the system was further developed by [William Rowan] Hamilton and [Carl Gustav Jacob] Jacobi into forms in which the equations of motion of a system represent the conditions for a stationary value of an integral of a single function.[22] The extension by [Edward] Routh and [Hermann von] Helmholtz to the case in which "ignored coordinates" are taken into account, was a long step in the direction of the desirable unification which would be obtained if the notion of potential energy were removed by means of its interpretation as dependent upon the kinetic energy of concealed motions included in the dynamical system.[23] The whole scheme of abstract Dynamics thus developed upon the basis of Lagrange's work has been of immense value in theoretical Physics, and particularly in statistical Mechanics, which is now a subject of enormous importance.

But the most striking use of Lagrange's conception of generalized coordinates was made by Clerk Maxwell, who in this order of ideas, and inspired on the physical side by the work of [Michael] Faraday, conceived and developed his dynamical theory of the Electromagnetic field, and obtained his celebrated equations. The

---

22 William Rowan Hamilton, "On a General Method in Dynamics," *Philosophical Transactions*, Vol. 124, Part I (Royal Society, 1834), pp. 247-308, and "Second Essay on a General Method in Dynamics," *Philosophical Transaction*, Vol. 125, Part I (Royal Society, 1835), pp. 95-144. Jacobi's memoir on the subject is excerpted and translated by Isaac Todhunter, *A History of the Progress of the Calculus of Variations during the Nineteenth Century* (Macmillan and Co., 1861), pp. 243-254.

23 Edward John Routh, *The Elementary Part of a Treatise on the Dynamics of a System of Rigid Bodies*, Seventh Edition (Macmillan and Co., 1905), Chapter VIII. Hermann von Helmholtz, "Principien der Statik monocyclischer Systeme," *Journal für die reine und angewandte Mathematik*, Volume 97 (1884), pp. 111-140.

form of Maxwell's equations enabled him to perceive that oscillations could be propagated in the electromagnetic field with the velocity of light, and suggested to him the Electromagnetic theory of light.[24] Heinrich Hertz, under the direct inspiration of Maxwell's ideas, demonstrated the possibility of setting up electromagnetic waves differing from those of light only in respect of their enormously greater length.[25] We thus see that Lagrange's work, conceived in the spirit he has himself described, was an essential link in a chain of investigation of which one result is of the kind which gladdens the heart of the practical man, viz. wireless telegraphy.

The habits of thought of such men as Lagrange have, however, their drawbacks, as I proceed to illustrate by means of a quotation from [William] Thomson [also known as Lord Kelvin] and [Peter] Tait's *Natural Philosophy*. With reference to the determination of the motions of a system disturbed from a position of equilibrium, by means of an equation of which the roots are the frequencies of the resulting vibrations, they write:

> It is remarkable that both Lagrange and Laplace fell into the error of supposing that equality among roots necessarily implies terms in the solution of the form $te^{\lambda t}$ (or $t \cos pt$), and therefore that for stability the roots must all be unequal. This we find in the *Mécanique Analytique*, [Seconde Partie, section VI. Art. 7 of] the second edition of 1811 published

24 James Clerk Maxwell, *A Treatise on Electricity and Magnetism* (Clarendon Press, 1873), Part IV.

25 Hertz, Heinrich. *Electric Waves*, translated by D. E. Jones (Macmillan and Company, 1893).

> three years before Lagrange's death, and repeated without change in the posthumous edition of 1853...
>
> It would be curious if such an error had remained for twenty-three years in Lagrange's mind. It could scarcely have existed even during the writing and printing of the Article for his last edition if he had been in the habit of considering particular applications of his splendid analytical work: if he had he would have seen that a proposition which asserted that the equilibrium of a particle in the bottom of a frictionless bowl is unstable if the bowl be a figure of revolution with its axis vertical, cannot be true.[26]

The correct theory of equal roots was first given by Routh.[27] That even physical instinct of the highest order does not always preserve the mind from such errors is curiously illustrated by the statement made a little further on by Thomson and Tait, that:

> [a]n error the converse of that of Laplace and Lagrange occurred in page 278 of our First Edition where it was said that: "Cases in which there are equal roots leave a corresponding number of degrees of indeterminateness in the ratios $l : m$, $l : n$, etc., and so allow the requisite number of arbitrary constants to be made up," without limiting this statement to the case of conservative positional and

26 William Thomson and Peter Guthrie Tait, *Treatise on Natural Philosophy*, Vol. I, Part I (Cambridge University Press, 1879), pp. 381-382.

27 Routh, *A Treatise on the Stability of Motion* (Macmillan and Co., 1877), pp. 5-9.

no motional forces, for which its truth is obvious at first sight....[28]

An interesting instance in which some very abstract departments of Mathematics were largely dependent for their origin upon concrete physical problems is the case of the problems connected with the vibration of strings and bars. These problems were discussed by [Daniel] Bernoulli, [Leonhard] Euler, and Lagrange, in the eighteenth century, and led to the attempt to represent arbitrarily prescribed functions by means of trigonometrical series.[29] To the discussions as to the possibilities in this direction are largely owing the progressive development of the notion of a Mathematical function, culminating in the extreme generality of the notion as it is at present accepted. In the hands of [Joseph] Fourier, the representation by trigonometrical series acquired the form which has become of so much importance in the resolution of disturbances, such as the tides, into harmonic components.[30] But for those of enquiring mind it was also the starting point of a long series of purely Mathematical

28 Thomson and Tait, p. 382. The "Cases" quotation occurs in the first edition (1867), p. 278.

29 Isaac Todhunter, *A History of the Theory of Elasticity and of the Strength of Materials*, Vol. I, edited by Karl Pearson (Cambridge University Press, 1886). For a chronicle of the Bernoulli and Euler work (including some correspondence between the two), see pp. 29-42, and for Lagrange, see pp. 58-67.

30 Joseph Fourier, *The Analytical Theory of Heat*, translated by Alexander Freeman (Cambridge University Press, 1878), originally *Théorie Analytique de la Chaleur* (Chez Firmin Didot, 1822). Early work was first published in "Mémoire sur la propagation de la Chaleur dans lees corps solides," *Nouveau Bulletin des Sciences* (la Société Philomathique, Mars 1808), pp. 112-116.

investigations, and it gave rise to all sorts of difficult questions of theoretic interest. The writings of [Peter Gustav Lejeune] Dirichlet, [Bernhard] Riemann, [Georg] Cantor, and more recently of [Henri] Lebesgue, are witnesses to the very general nature of the discussions which have arisen in this connection; and which, to a student of the history of the Science, appear as the natural evolution in an order of ideas which had its origin in the problems of vibrations and of the conduction of heat.[31] The developments which arose in this way have been a dominant factor in the modern transformation of Mathematical Analysis.

31 The contributions of Dirichlet, Riemann, and Cantor are discussed in Philip Jourdain's introduction to Georg Cantor, *Contributions to the Founding of the Theory of Transfinite Numbers* (Open Court Publishing Company, 1915). See also Henri Lebesgue, "Intégrale, Longueur, Aire," *Annali di Matematica*, Serie III, Tomo VII (1902), pp. 231-359.

VII

[Henri] Poincaré has said that Mathematics is the art of giving the same name to different things.[32] We have here in epigrammatic form the statement of a most fundamental characteristic of modern Mathematics; and one which explains the high degree of abstractness that many of its branches exhibit. It has frequently happened that processes which have been employed at different times and by different investigators in connection with things of very diverging characters have been perceived to have an underlying identity that was in the earlier stages quite unsuspected. When this perception has once been made, these processes dealing with things that are superficially of the most varying characters become, from the abstract point of view of the Mathematician, identical; the same names and the same language become applicable to all of them. For example, the modern theory of groups of operations is of this character.

Whatever be the more concrete character of a set of operations, from the point of view of the theory of groups, the group

32 "Je ne sais si je n'ai pas déjà dit quelque part que la Mathématique est l'art de donner le même nom a des choses différentes." Henri Poincaré, "L'Avenir des Mathématiques," *Bulletin des Sciences Mathématiques*, Tome XXXII (Gauthier Villars, 1908), pp. 168-190; p. 174.

alone is the one essential characteristic of these operations, and all sets of operations, whatever be their nature, which have the same group, may be regarded as identical. The analysis of the nature of the abstract group itself furnishes results which are immediately interpretable for operations of the most varied character, provided they have one and the same group. The relation between operations connected with the roots of a quintic [fifth power] equation and the operations connected with the rotations of an icosahedron [20-sided] into coincidence with itself is an example of this representation by one and the same abstract group.[33]

The many methods of transformation by means of which a single theorem once established is capable of translation so as to give a whole series of theorems about objects that are in their nature quite distinct from those with which the original theorem had to do, afford examples of the identity of Mathematical form underlying differences in every aspect except just that one which is of importance for the specific purpose of the Mathematician. For him the most striking differences between different species of objects may be all irrelevant when he is concerned only with certain aspects of those objects which are identical in the different species for his special purpose. The importance which the Mathematician attaches to elegance and conciseness in his results and formulae is not due merely to a species of aesthetic taste, although such aesthetic instinct often acts as an efficient guide. By attention to these details of Mathematical form he is frequently led to the discovery of unsuspected relations between different departments, and to a synthesis

33 Felix Klein, *Lectures on the Ikosahedron, and the Solution of Equations of the Fifth Degree*, translated by George Gavin Morrice (Trübner & Co., 1888).

in which unity takes the place of an earlier diversity, leading to a true economy of thought.

An examination of the mode in which those vast modern developments which we group under the name of Geometry have arisen, illustrates in an instructive manner the natural growth of Mathematical theories from the parent stem rooted in empirical observation of the physical world, and it brings to light the inherent necessity which causes such a branch of knowledge to take up forms of ever-increasing grades of abstractness. In its origin, as a study of the forms and relative positions of material bodies, Geometry is a physical Science built up by observation and experiment; on its practical side, as applied Geometry, it still retains that character. But at an early stage, with the Greeks, it became a purely rational Science, for the further development of which empirical observation was no longer a necessity.

In rational Geometry, the objects dealt with, points, straight lines, circles, conics, etc., are not physical objects, but are abstractions from objects perceived by the senses; they are idealizations of such objects of perception, in that they are regarded as possessing in perfection certain properties which we discern in the corresponding physical objects as only approximately realized in greater or less degree. Thus, for example, the straight line in rational Geometry has in absolute perfection the properties of linearity and straightness, whereas these properties are only imperfectly realized in any physical object which we may take as affording a sensuous image of a straight line. The precise properties of these idealized objects are fixed by means of some scheme of definitions, axioms, and postulates, the nature of which is to a large extent, but as we shall see not wholly, determined by empirical observation of actual relations in the physical domain. Only on the condition that sense data underwent this process of idealization could Geometry as an exact rational Science become possible.

The procedure in other physical sciences is not generically different from that in Geometry. Any one of them becomes a rational Science when, and so far as, a schematic representation of the phenomena with which it deals is set up. The peculiarity of Geometry is that it became a purely rational Science earlier, and by more rapid stages, than could be the case in other departments of physical investigation which have not yet emerged from the stage in which empirical observations form an essential element in the process of furthering our knowledge. The reason of the fact that Geometry, having become a purely rational Science, is in a more advanced stage than other Sciences, is to be found in the comparative obviousness of the empirical observations on which it is based.

The history of the development of Geometry is of general interest because Geometry may be regarded as the type to which every Science may be expected to conform at the distant time when it has become completely rational. When that distant consummation has been reached, laboratories will be useful only for illustrative and didactic purposes, just as drawings and models are still used in Geometry, but will no longer be essential for purposes of research; the progress of such a Science will, as is the case for Geometry at present, consist of a detailed development of the schematic representations of the phenomena. Just as Geometry has no need of further empirical fact, completely rationalized Physics and Chemistry, as ideal schemes, would contain within themselves every element which could be supplied by physical observation, and would no longer be dependent for their further progress upon the work of the experimenter. It is of course true that the amount and nature of the empirical data required by such Sciences as Physics and Chemistry are of such a character that we can hardly conceive such a consummation as the one I have spoken of as anything but an ideal; how far that ideal is capable of realization posterity alone will decide. For a long time to come these Sciences can be only partly rational, but their line of progress is such that the rational part of them must ever increase relatively to their empirical part.

VIII

I HAVE SAID that the properties and relations assigned to the ideal objects of rational Geometry arose out of suggestions originating in observations of spatial properties and relations in the physical world; there are however noteworthy limitations to the amount of leading that can in this regard be obtained from our actual perceptions. In the first place, all our spatial perceptions are affected by an essential element of inexactitude, the amount of which varies with the precision of the instruments we use for measurements, but which can never be wholly eliminated. In the second place, all our actual observations have reference to some more or less bounded, and certainly finite, portion of what we call physical space, although we have the intuition that it is always possible to pass beyond the boundaries of the space at any time observed. In rational Geometry, we have in the scheme of axioms, definitions, and postulates, not only to make precise statements as to the possession by certain entities of certain precise properties, but we have also to make statements which refer to what happens in every part of unbounded space; and to do this we have to pass in certain respects beyond anything we can learn from physical observation.

In one point, namely in connection with what is usually known as the axiom of parallels, this disability has proved to be of the greatest importance in the development of the Science. Euclid's axiom of parallels, being a statement of what happens in unbounded space, is essentially incapable of direct verification.[34] Indirect verification, for example, by observation of the sum of the angles of a triangle is indecisive, on account of the essential inexactitude of our measurements, and on account of the fact that the size of the triangles that can be observed even by the Astronomer is limited. All attempts to prove the truth of the axiom, that is to deduce it from the other axioms and postulates, proved a failure in face of the most determined efforts to throw light upon the matter; it has now been demonstrated that this failure was inevitable. It is absolutely necessary for the development of rational Geometry either to postulate, as Euclid does, in some form or other, the truth of the axiom of parallels, or else to substitute for it some other postulate of a divergent character, but still not inconsistent with our intuitions of actual spatial relations.

During the last century, the actual failure to prove the truth of Euclid's axiom within the rational scheme, led to the investigation of the results of making a different postulation, and at least two systems, that of Bolyai-Lobachevsky [1829-31], and that of [Bernhard] Riemann [in 1854], were invented.[35] Each of these contains a

34 "[Let the following be postulated:] That, if a straight line falling on two straight lines make the interior angles on the same side less than two right angles, the two straight lines, if produced indefinitely, meet on that side on which are the angles less than the two right angles." Euclid, *Elements*, Book I, Postulate 5.

35 This history is briefly recounted in David Eugene Smith, *History of Mathematics*, Vol. II (Ginn and Company, 1925), pp. 335-338. János Bolyai, "Appendix: Scientiam spatii absolute veam exhibens," in Farkas Bolyai, *Tentamen juventutem studiosam in elementa matheseos purae, elementaris ac sublimioris,*

substitute for Euclid's axiom, and is inconsistent with it, and they are inconsistent with one another; but each is, when suitably interpreted, a sufficient representation of our actual space perceptions, although not so simple a representation as that of the Euclidean scheme.[36] It has been demonstrated that each of these schemes is logically self-consistent, and thus the possibility of diverging conceptual schemes for the representation of a single set of perceptual data has been established. This is a result of the highest possible general interest. The view is no longer tenable that the relations of time and space exist in the mind as empty forms of thought, as in [Immanuel] Kant's scheme, prior to all perception, and as forms in which all our spatial and temporal perceptions clothe themselves.[37]

---

methodo intuitiva, evidentiaque huic propria, Vol. I (Maros Vásárhelyini, 1832). For English, see John Bolyai, "The Science Absolute of Space," translated by George Bruce Halsted, *Scientiæ Baccalaureus*, Vol. I, No. 4 (June 1891). Nicholaus Lobatschewsky, *Geometrical Researches on the Theory of Parallels* [1840], translated by George Bruce Halsted (University of Texas, 1891). Bernhard Riemann, "On the Hypotheses Which Lie at the Bases of Geometry [1854]," translated by William Kingdon Clifford, *Nature*, Vol. VIII, No. 183 (May 1, 1873), pp. 14-17, and No. 184 (May 8, 1873), pp. 36-37.

36 "The axiom of parallels may be deviated from in two directions. In the one [developed by Riemann] it is supposed that every two lines in the plane must meet; no line parallel to another can be drawn through the same point in the lane. Deviating in the other [Bolyai-Lobachevsky] direction we have several lines drawn through the point which never meet the given line; they diverge from it as lines on an [sic] hyperboloid may diverge." Simon Newcomb, "The Philosophy of Hyper-Space," *Science,* N.S. Vol. VII. No. 158 (January 7, 1898), pp. 1-7; p. 6. In other words, Riemann bans parallel lines and Bolyai-Lobachevsky makes every line parallel.

37 "It is therefore beyond the reach of doubt, and not possible only or probable, that space and time, as the necessary conditions of all experience,

Nor is it true that the conceptual schemes for spatial and temporal relations are simple products of empirical observation. They could never have arisen apart from our actual experience of the external world, but they contain much more than the raw data obtained from perception. The rise from the rough data of sensuous perception to a rational scheme representing the relations involved can only be accomplished by a process in which the reflective activities of the mind contribute an essential element which is lacking in the data themselves.

I have now to refer to a most important fact in this connection, namely that the mode in which the data of the senses can be idealized and completed leaves a latitude in which there is a large element of free choice left to the mind. In the case of Geometry this element of free choice has been exercised in the construction of various types of Geometry, all of which are logically possible, and several of which are equally applicable for the representation of our actual spatial experience. Thus Geometry has been transformed into Geometries. This remarkable result suggests that in other Sciences it is possible that the system of laws forming the schematic representation of the phenomena may not necessarily be unique; that there may be in any given domain several or even an unlimited number of different rational schemes which will all be equally valid; although, as in the case of Geometry, they will not be all equally simple.

The great discovery of [René] Descartes which in our time has become part of the mental furniture of multitudes of

external and internal, are purely subjective conditions of our intuition…." Immanuel Kant, *Critique of Pure Reason*, translated by Friedrich Max Müller (Macmillan Company, 1896), p. 39.

non-Mathematicians, in its various applications to graphical representation of variables, gave rise to a revolutionary increase in the scope and in the power of geometrical Science, by reducing Geometrical investigation to a form in which the relations of number are made to do the work, and in which the powerful and expanding methods of Algebraical Analysis are made available for research in the domains of the various Geometries.[38] Here again the mode in which geometrical objects shall be correlated with numbers and number-systems leaves to the mind a large element of free choice, and has led to the study of schemes varying in accordance with the manner in which this free choice is exercised. The correlation of the points of a straight line, for example, with a scheme of numbers, leaves latitude of the kind I have described; we have, for example, not only the more usual schemes of correlation in which the existence of actually infinitesimal segments of a straight line is denied, in the so-called Archimedean systems, but there have been developed also non-Archimedean systems in which quite different postulations are fundamental.[39] The differences in these various modes of correlation are in a region which is inaccessible to any kind of verification by sense-perceptions, and is of no direct interest to the Physicist, but the contemplation of the various possibilities implied in these schemes is forced upon the mind of the Mathematician in the course of his endeavours to explore and complete the rational foundations of the subjects which he studies.

38 *The Geometry of René Descartes*, translated by David Eugene Smith and Marcia L. Latham (Open Court Publishing Company, 1925).

39 David Hilbert, *The Foundations of Geometry*, translated by E. J. Townshend (Open Court Publishing Company, 1902).

IX

Time will not allow me to refer in detail to the vast modern developments such as projective Geometry, descriptive Geometry, line Geometry, etc.,[40] but I will refer briefly to one further great step in abstraction or generalization. All geometrical schemes may be regarded as specializations of an abstract scheme in which the subject-matter consists of the possible determinations which can be made within an ordered manifold, in which the order is not necessarily spatial or temporal, but in which the notion of order is purely abstract, and of which spatial and temporal orders are merely specializations. The elements of such a manifold are regarded as purely ideal objects, identified only by means of some assumed correlation with a number-system which may have any number of dimensions, or even an infinite number of dimensions; the order of the elements is assigned by means of the ordinal relations of the number-systems with which the elements are correlated. A generalization of this kind would appear to be the final abstraction, of which all

40 David Eugene Smith suggests these developments comprise "the renaissance of pure geometry." David Eugene Smith, *History of Mathematics*, Vol. II (Ginn and Company, 1925), p. 331. Projective geometry is typically associated with Jean-Victor Poncelet, descriptive geometry with Gaspard Monge, and line geometry with Julius Plücker.

Geometries are specialized examples; from this final generalization all remnants of specially spatial or specially temporal intuition have been removed by abstraction, the relations of order alone remaining. The advantages of the contemplation of such a purely abstract scheme consist in its suitability as a means of investigation and classification of all those schemes of a less abstract character which may be subsumed under it as specialized instances. It may seem strange that the investigation of the relations within an ordered manifold of purely ideal objects should still be sometimes spoken of as Geometry, although the relations with which it deals have no necessary, but only an adventitious, connection with what we call spatial relations.

I have already referred to the difficulties which occur in connection with the extension of our intuitions of finite relations to the domain of the infinite. I propose here to give one illustration which I think exhibits strikingly the necessity for caution in this matter, the lack of which has often led to generalizations which were afterwards found to be untenable, at least in their original form.

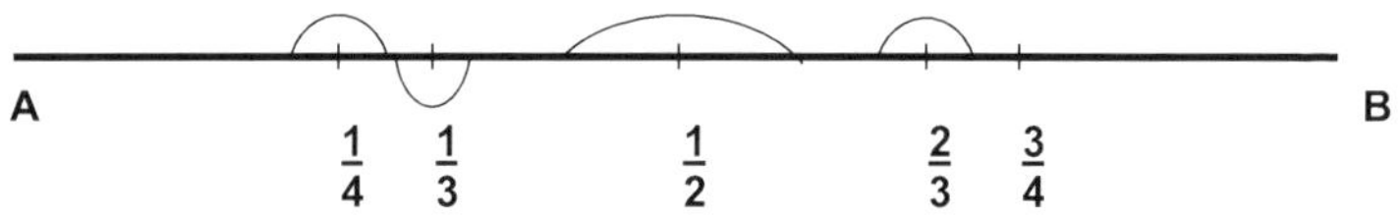

Let us contemplate a straight line *AB*, say a mile long. If we imagined that there were removed from this straight line a number of segments, or portions, the total length of which is less than one foot, what would remain? Naturally, a number of segments whose total length would be under a foot less than one mile. That is the answer given by our intuition, and it is a perfectly correct one, but only so long as the number of segments removed is finite. I propose to show you that it is possible to conceive an infinite set of segments removed from the straight line, the total length of which is less than one foot, in such wise, that in what remains there are no segments at all, but only points, so that nothing is left which can be said to have a length.

Take the various proper fractions

$$\frac{1}{2}, \quad \frac{1}{3}, \frac{2}{3}, \quad \frac{1}{4}, \frac{2}{4}, \frac{3}{4}, \quad \frac{1}{5}, \frac{2}{5}, \frac{3}{5}, \frac{4}{5}, \quad \frac{1}{6}, \frac{2}{6}, \frac{3}{6}, \frac{4}{6}, \frac{5}{6} \ldots.$$

Here every proper fraction occurs at a definite place in the sequence, and in fact it occurs not only once, but an indefinite number of times.

Let us suppose that, corresponding to each fraction, a point representing the corresponding fraction of the mile is taken on *AB*. Remove a segment, of length half a foot, with its centre at the point $\frac{1}{2}$. Then remove a segment, of length $\frac{1}{2^2}$ of a foot, with its center at the point $\frac{1}{3}$; then a segment, of length $\frac{1}{2^3}$ foot, with its centre at the point $\frac{2}{3}$; and so on. We thus imagine removed a segment, of length $\frac{1}{2^n}$ foot, with its centre at the point represented by the *n*th fraction. We can disregard those fractions which are not in their lowest terms, because the segments which have the corresponding points for centres have been already removed. The total length of all the segments which we conceive to have been removed is, since some of them are contained in others, or overlap others, certainly less than $\frac{1}{2}+\frac{1}{2^2}+\frac{1}{2^3}+\ldots+\frac{1}{2^n}+\ldots$ of a foot, and is therefore less than one foot.

All the points corresponding to the rational fractions exist on *AB*, and there can be drawn no length on *AB* which is devoid of these points. As we regard all these points as existing, we may also regard all the segments with these points as centres as existent, and to have been removed from *AB*. When these segments have all been removed there can be no segment of *AB* left, because any such segment contains points of the above set, and thus segments with such points as centres will have been removed. Nothing can remain in *AB* except points which make up no single segment.

We have thus the apparently paradoxical result that, by removing from a straight line of length one mile, a set of portions, the total length of which is less than one foot, there are left no portions of the straight line at all, but only single points.

X

Mathematics may perhaps be compared with a set of games which Mathematicians occupy themselves in playing. In all the earlier games of the set the rules were suggested by physical experience, and were useful as a representation of the behaviour of actual bodies; like other species of games these were imitative in character. But it was found necessary to give the rules complete precision in their form, and it ultimately turned out that this could be done in a variety of ways, and hence what was originally one game split up into a variety of different games played according to different rules. Those who were primarily interested in the games for their own sake were interested not only in those with the simplest sets of rules, but in others of less simplicity. Those, viz. the Physicists, who are primarily interested only in the imitative aspect of the games, confine their attention to those species of games in which the rules are of the simplest possible character compatible with their imitative purpose.

To the Mathematician it has always been a matter of supreme interest to find out what would be the effect of altering or extending the rules of a game in various ways, thus creating new games. Every now and then it has been discovered that what appeared to be quite different games with different sets of rules could be reduced to a single game with a single set of rules which in a sense superseded the original games. The only rules that are absolutely essential in all the games are those known as the laws of logic; the other rules always possess a greater or less degree of arbitrariness. When the players become too skilful, and the games become too complicated, onlookers have sometimes a difficulty in comprehending the ardour of the players, and are apt to think that the matter has been carried beyond all reason. But the Mathematical games are played with definite purposes; they are infinite in variety; they call forth and develop some of the highest faculties of the human mind. A true estimate of their value can be made only if they are observed in a sympathetic spirit, and in the light of a comprehensive view of the history of human thought. If this is done, the general verdict cannot fail to be a favourable one.

I have endeavoured to point out, in a necessarily fragmentary and incomplete manner, some of the main differences which distinguish the view of Mathematics taken by the Mathematician from its aspect for the Physicist. In conclusion, I would emphasize the fact that, although the study of Mathematics which the Physicist or the Engineer must make has rightly many points of difference from the study as pursued for other purposes by the Mathematician proper, it must nevertheless be of a serious and thorough character, if it is to be really effectual for the purposes for which it is needed. Many branches of modern Mathematics may and must be disregarded by those who study Mathematics mainly with a view to its application in Physical Science, and many details may be omitted which are of subordinate importance; but so far as it is studied at all it must be studied as a Science, not as an Art. Its study must be pursued in a spirit of patience, without too exclusive attention to those parts of it which can be immediately applied to practical problems. The instruments provided by Mathematics are of so delicate a character that they are apt to break in the necessarily unskilful hands of those that do not understand their construction.

Classical and Modern Differential Geometries: A Response Essay

J. Brian Pitts

University of Lincoln, University of Cambridge, University of South Carolina

Professor Hobson's address remains, over a century later, a useful and fair description of the relations of mathematics and physics, except that one might now wish to distinguish descriptive and normative versions of his view. By now one could give additional examples of how physics provides problems that inspire the mathematician's reflection, which proceeds to develop to greater levels of abstraction with no immediate physical application; one could also give examples of how in some cases a surprising physical application eventually was made of mathematics thus already developed. Writing as one now does (unlike Hobson) after Bourbaki and related attitudes have affected certain branches of mathematics, in which notions of elegance have gained ground at the expense of comprehensiveness (Stewart, 1995), one might suggest that normatively his view still applies, but descriptively, mathematics has developed a novel sort of independent streak beyond what Hobson discussed, and of which he might have been wary. As he noted, pure mathematicians tend to find the well-behaved functions, converging Taylor and Fourier series, *etc.*, boring, though these cases are frequently those most relevant to physics. That older pure mathematician's attitude did not come at the expense of comprehensiveness, however, as it increasingly came to do in the 1930s with Bourbaki; then ugly results did not need to be mentioned at all, and one was free to develop the field in a way that made them nearly inconceivable, leaving only the pretty results. This movement led to a breakdown in communication with physicists (Stewart, 1995). Hobson's remarks about the pure mathematicians' comments about Bessel functions and spherical harmonics suggest that he would have recognized excess in Bourbaki along these lines. Differential geometry being the field of mathematics known best to me, I will largely confine my remarks to it.

As modern differential geometry has increasingly privileged aesthetic criteria notationally and conceptually, generations of students have had reduced contact with the generous collection of geometric objects (things like tensors) that one used to find more commonly in both mathematics and physics books (Schouten, 1954; Anderson, 1967). These include tensor densities of arbitrary (including non-integral) weight and various oriented *vs.* non-oriented quantities including polar *vs.* axial vectors and scalars *vs.* pseudoscalars. If such less famous quantities played no useful role in calculations, then the modern mathematical quest for elegance would benefit the working physicist as well. Is the modern style expressively adequate for local problems?

In fact such classical and now neglected entities as Schouten and Anderson displayed are useful, probably important and perhaps essential in some cases. Flanders's book on differential forms "never [takes] absolute values of Jacobians." (Flanders, 1989, p. 1). By contrast William Burke remarked:

> Although most of the material in this book is quite standard, twisted tensors are not.. . . Most of the geometric objects that we will need in our physical theories are actually twisted tensors.. . . We could make out a good case that the usual differential forms are actually the twisted ones,. . . (Burke, 1985, p. 183)

Axial vectors and pseudoscalars are common in particle physics. In certain physics circles involving General Relativity and gravitation one hears in conversation that one "cannot really calculate anything" using coordinate-free notation. This is doubtless an exaggeration, depending on how one construes "anything"; indeed there are things that one can calculate more easily or at least is more likely to think up with modern geometric ideas than with classical ones, some of which are relevant to local problems. But it still seems to be true that attempting to satisfy modern coordinate-free strictures thus far has a retarding tendency for local problems, because the quantities most appropriate for certain problems either do not admit coordinate-free formulations, or have not yet been given coordinate-free formulations (with no apparent urgency to fill the gap), or admit coordinate-free formulations only in a Pickwickian sense of feeding in a basis (not necessarily natural) or the like in order to get a *number(s)* that one can then manipulate more fully (such as taking a power or root), or have been given coordinate-free formulations that remain poorly known despite the passage of decades and/or are surprisingly difficult (Woodhouse, 1992) in comparison to either the elegance of coordinate-free tensors-in-themselves $g = g_{\mu\nu}\, dx^{\mu} \otimes dx^{\nu}$ or the conceptual unity and simplicity of geometric objects as collections of components with a transformation rule along the lines of $g'_{\alpha\beta} = g_{\mu\nu} \frac{\partial x^{\mu}}{\partial x^{\alpha'}} \frac{\partial x^{\nu}}{\partial x^{\beta'}}$. Anthony Zee recounts an illuminating story.

> Coordinate-free notations are great for proving general theorems but are not so good for calculating. . . . [A leading young researcher at the gravity-particle physics interface had during] grad school, to deepen his understanding of Einstein gravity,. . . enrolled in a course taught by a famous mathematician. As it happened, he was the only student able to do the problems in the final exam involving actual calculations: he did them by first using old fashioned indices and then translating back into the abstract notation used in the course. (Zee, 2013, p. xxii)

Workers in loop quantum gravity feel free to use tensor densities in the classical way, because sometimes a density of a certain weight precisely captures what the physics calls for and apparently nothing else does. Presumably they read old books or papers written by old people to learn these tools. To give another example, in the context of perturbative quantum gravity Bryce DeWitt found that a metric densitized to weight $-\frac{5}{11}$ or inverse metric densitized to weight $\frac{5}{9}$ turned an infinite series into a polynomial (DeWitt, 1967), a distinction potentially of importance in field theory (though ultimately not so crucial in this case). How many slots does a density of weight $\frac{5}{9}$ have? How about a tensor density of weight $\sqrt{2}$? Densities of arbitrary real weight naturally suggest different mass terms for massive spin 2 gravity (Ogievetsky and Polubarinov, 1965) and so are helpful, if not important, in physical theory construction. Thus the working physicist, privileging the expressive power needed for calculation and theory construction, will want answers to such questions.

A related problem, though not entirely distinct, involves surplus structure. It is a commonplace that getting rid of surplus structure in physical theories is a good idea; Lorentz's ether is a standard example that people love to hate. But surplus structure has a way of showing up in places that one might not expect, especially if one has modern mathematical tastes. For example, one sees the use of the Hodge dual when the volume-free non-oriented Levi-Civita dual (using the purely numerical ± 1, 0 ϵ antisymmetric symbol) is conceptually more appropriate and topologically more robust. One also sees conformal Killing vector fields defined using surplus structure. A lean definition in terms of conformal geometry, the geometry of the conformal metric tensor, a density of weight $-\frac{2}{n}$ in n dimensions (Thomas, 1925), with determinant ± 1, is

$$\pounds_{\xi}\hat{g}_{\mu\nu} = 0$$

(Yano, 1955, p. 32), cutting nature at the joints. Thomas's work was largely forgotten, however (Branson, 2005). One typically sees conformal Killing vector fields defined as

$$\pounds_{\xi}g_{\mu\nu} = f g_{\mu\nu}$$

for some physically meaningless function f along with the irrelevant volume element $\sqrt{|g|}$ in the metric. (A metric can be decomposed into irreducible parts as

$$g_{\mu\nu} = \hat{g}_{\mu\nu} \sqrt[n]{|g|},$$

permitting separate treatment of the conformal metric density and the volume.) That is akin to translating a sentence from one language to a related language (perhaps German to Dutch, akin to a small coordinate transformation) in cases where the words in the predicate (analogous to the conformal metric density) are all cognates (akin to a conformal Killing vector) while the subject words (analogous to a volume element) might differ radically. If the predicates (when left unchanged by a small transformation) are the topic of interest, then why not just translate the predicates and omit the subjects? Mathematical existence is cheap, so having an irrelevant volume element in $g_{\mu\nu}$ and an irrelevant function f does not harm the mathematician's interests, but the physicist's calculations and the philosopher's reflections are both impeded by surplus structure. There is also a physical bias involved. Someone working on, say, Brans-Dicke gravity, in which there is one conformal metric density along with two volume elements, should be able to define conformal Killing vector fields, which are shared, without having to pick the Jordan 'frame', the Einstein 'frame', or indeed any volume element. Expressing things in terms of a metric, as though existence and uniqueness of a volume element can be taken for granted, reflects and reinforces a bias for General Relativity and against other theories.

The relationship of spinors to coordinate systems is another interesting example of surplus structure—one quite important in efforts to figure out what one might or should mean by absolute space-time structure in the tradition of James Anderson's analysis (Anderson, 1967; Pitts, 2006). Weyl concluded in 1929 that spinors could not be spinorial with respect to coordinates (Weyl, 1929); E. Cartan concluded a book on spinors with a similar result (Cartan and Mercier, 1966, p. 151). Thus one needed to couple spinors to an orthonormal basis of vectors, introducing an extra 6 components (in 4 dimensions) and 6 extra bits of local Lorentz gauge freedom to deprive them of physical meaning. While Weyl's mathematical result is fine, his glosses shows a lack of imagination in giving up some tacit premises. One can show, as was largely achieved constructively by Bryce DeWitt (Seligman [DeWitt], 1949; DeWitt, 1950) and by Ogievetksy and Polubarinov (Ogievetski˘i and Polubarinov, 1965) with greater clarity, both using perturbative expansions, that spinors form almost part of a geometric object with the metric, though not by themselves (Pitts, 2012); one has, nearly, a nonlinear realization of general coordinate transformations that is linear for the 15-parameter conformal group. Such ideas for coupling spinors to the (principal square root of the) metric remain poorly known well over half a century latter (Kay, 2020), though they are frequently partly reinvented by the back door *via* imposing a symmetric tetrad gauge condition (*e.g.*, (van Nieuwenhuizen, 1981; Woodard, 1984)). Occasionally such things are contemplated in a modern geometric style (Bourguignon and Gauduchon, 1992). The history of this issue might be a good test case for sociologists of mathematics and physics.

It would seem that modern differential geometry, while making remarkable progress along the directions that it has set for itself, has not yet convinced a lot of working physicists that it is an asset for their calculations and derivations using local methods. It might be useful to see an effort to extend the modern style of differential geometry to see how far it can recover the expressive capacity of classical differential geometry, while of course retaining its own extant virtues. Such a quest would permit a just evaluation of the strengths and weaknesses of each project and could potentially reduce the split between mathematicians and working physicists regarding differential geometry. (One might see an interesting example of convergence between modern and classical geometry in a recent paper on the hole argument, arguing that, as was already accepted by Einstein using classical differential geometry, a passive rather than active interpretation is preferable (Weatherall, 2018).) In that case the divergence between the normative sense of Hobson's view and the descriptive sense (which has fallen somewhat into disrepair) would also be removed, at least regarding differential geometry. I am not equipped to comment so much on other areas of mathematics and their physical applications.

References

Anderson, J. L. (1967). *Principles of Relativity Physics*. Academic, New York.

Bourguignon, J.-P. and Gauduchon, P. (1992). Spineurs, op´erateurs de Dirac et variations de m´etriques. *Communications in Mathematical Physics*, 144:581–599.

Branson, T. (2005). Conformal structure and spin geometry. In Bourguignon, J.-P., Branson, T., Chamseddine, A., Hijazi, O., and Stanton, R. J., editors, *Dirac Operators: Yesterday and Today: Proceedings of the Summer School and Workshop, CAMS-AUB, Lebanon, August 27-September 7, 2001*, pages 163–191. International Press, Somerville, Massachusetts.

Burke, W. L. (1985). *Applied Differential Geometry*. Cambridge University Press, Cambridge.

Cartan, E. and Mercier, A. (1966). *The Theory of Spinors*. Massachusetts Institute of Technology Press, Cambridge. French original 1937.

DeWitt, B. S. (1950). On the application of quantum perturbation theory to gravitational interactions. https://repositories.lib.utexas.edu/handle/2152/9620.

DeWitt, B. S. (1967). Quantum theory of gravity. II. The manifestly covariant theory. *Physical Review*, 162:1195–1239.

Flanders, H. (1989). *Differential Forms with Applications to the Physical Sciences*. Dover, Mineola, New York. Corrected reprint from Academic Press (1963), New York.

Kay, B. S. (2020). Editorial note to: Erwin Schr¨odinger, "Dirac electron in the gravitational field I". *General Relativity and Gravitation*, 52. arXiv.org/1906.10765, with addendum with Gary Gibbons.

Ogievetskiˇi, V. I. and Polubarinov, I. V. (1965). Spinors in gravitation theory. *Soviet Physics JETP*, 21:1093–1100.

Ogievetsky, V. I. and Polubarinov, I. V. (1965). Interacting field of spin 2 and the Einstein equations. *Annals of Physics*, 35:167–208.

Pitts, J. B. (2006). Absolute objects and counterexamples: Jones-Geroch dust, Torretti constant curvature, tetrad-spinor, and scalar density. *Studies in History and Philosophy of Modern Physics*, 37:347–371. arXiv:gr-qc/0506102v4.

Pitts, J. B. (2012). The nontriviality of trivial general covariance: How electrons restrict 'time' coordinates, spinors (almost) fit into tensor calculus, and 7/16 of a tetrad is surplus structure. *Studies in History and Philosophy of Modern Physics*, 43:1–24. arXiv:1111.4586.

Schouten, J. A. (1954). *Ricci-Calculus: An Introduction to Tensor Analysis and Its Geometrical Applications.* Springer, Berlin, second edition. http://link.springer.com/book/10.1007

Seligman [DeWitt], C. B. (1949). *I. The Theory of Gravitational Interactions. II. The Interaction of Gravitation with Light.* PhD thesis, Harvard University. Supervisor Julian Schwinger.

Stewart, I. (1995). Bye-bye Bourbaki: Paradigm shifts in mathematics. *The Mathematical Gazette*, 79:496–498.

Thomas, T. Y. (1925). Invariants of relative quadratic differential forms. *Proceedings of the National Academy of Sciences of the United States of America*, 11:722–725.

van Nieuwenhuizen, P. (1981). Classical gauge fixing in quantum field theory. *Physical Review D*, 24:3315–3318.

Weatherall, J. O. (2018). Regarding the 'hole argument'. *The British Journal for the Philosophy of Science*, 69:329–350. arXiv:1412.0303 [physics.hist-ph].

Weyl, H. (1929). Gravitation and the electron. *Proceedings of the National Academy of Sciences of the United States of America*, 15:323–334.

Woodard, R. P. (1984). The vierbein is irrelevant in perturbation theory. *Physics Letters B*, 148:440–444.

Woodhouse, N. M. J. (1992). *Geometric Quantization*. Clarendon Press, Oxford, second edition.

Yano, K. (1955). *The Theory of Lie Derivatives and Its Applications*. North-Holland, Amsterdam.

Zee, A. (2013). *Einstein Gravity in a Nutshell*. Princeton University Press, Princeton.

Appendix: Excerpted Writings of E. W. Hobson

I. The following excerpts are drawn from Hobson's address[41] before the British Association for the Advancement of Science on September 1, 1910, in Sheffield, England.

Probably no other department of knowledge plays a larger part outside its own narrower domain than Mathematics. Some of its more elementary conceptions and methods have become part of the common heritage of our civilisation, interwoven in the everyday life of the people. Perhaps the greatest labour-saving invention that the world has seen belongs to the formal side of Mathematics; I allude to our system of numerical notation. This system which, when scrutinized, affords the simplest illustration of Mathematical form, has become so much an indispensable part of our mental furniture that some effort is required to realise that an apparently so obvious idea embodies a great invention; one to which the Greeks, with their unsurpassed capacity for abstract thinking, never attained. An attempt to do a multiplication sum in Roman numerals is perhaps the readiest road to an appreciation of the advantages of this great invention. In a large group of sciences, the formal element,

41 E. W. Hobson, Presidential Address to Section A, *Report of the Eightieth Meeting* (Sheffield), British Association for the Advancement of Science (John Murray, 1911), pp. 509-522. Large blocks of text have been broken up into smaller paragraphs for ease of reading.

the common language, so to speak, is supplied by Mathematics; the range of the application of mathematical methods and symbolism is ever increasing.

Without taking too literally the celebrated dictum of the great philosopher Kant, that the amount of real science to be found in any special subject is the amount of Mathematics contained therein,[42] it must be admitted that each branch of science which is concerned with natural phenomena, when it has reached a certain stage of development, becomes accessible to, and has need of, mathematical methods and language; this stage has, for example, been reached in our time by parts of the science of Chemistry. Even Biology and Economics have begun to require mathematical methods, at least on their statistical side, As a science emerges from the stages in which it consists solely of more or less systematised descriptions of the phenomena with which it is concerned in their more superficial aspect: when the intensive magnitudes discerned in the phenomena become representable as extensive magnitudes, then is the beginning of the application of mathematical modes of thought; at a still later stage, when the phenomena become accessible to dynamical treatment, Mathematics is applicable to the subject to a still greater extent.

Mathematics shares with the closely allied subject of Astronomy the honour of being the oldest of the sciences. When we consider

42 Hobson is referring here to Kant's Preface to *Metaphysical Foundations of Natural Science* (1786). Original German: "...so wird Naturlehre nur so viel eigentliche Wissenschaft enthalten, als Mathematik in ihr angewandt werden kann." Translated by Ernest Belfort Bax: "...a doctrine of nature can only contain so much science proper as there is in it of applied mathematics."

that it embodies, in an abstract form, some of the more obvious, and yet fundamental, aspects of our experience of the external world, this is not altogether surprising. The comparatively high degree of development which, as recent historical discoveries have disclosed, it had attained amongst the Babylonians more than five thousand years B.C., may well astonish us. These times must have been preceded by still earlier ages in which the mental evolution of man led him to the use of the tally, and of simple modes of measurement, long before the notions of number and of magnitude appeared in an explicit form.

I have said that Mathematics is the oldest of the sciences; a glance at its more recent history will show that it has the energy of perpetual youth. The output of contributions to the advance of the science during the last century and more has been so enormous that it is difficult to say whether pride in the greatness of achievement in his subject, or despair at his inability to cope with the multiplicity of its detailed developments, should be the dominant feeling of the mathematician…

All attempts to characterise the domain of Mathematics by means of a formal definition which shall not only be complete, but which shall also rigidly mark off that domain from the adjacent provinces of Formal Logic on the one side and of Physical Science on the other side, are almost certain to meet with but doubtful success; such success as they may attain will probably be only transient, in view of the power which the science has always shown of constantly extending its borders in unforeseen directions. Such definitions, many of which have been advanced, are apt to err by excess or defect, and often contain distinct traces of the personal predilections of those who formulate them…

The belief is very general amongst instructed persons that the truths of Mathematics have absolute certainty, or at least that there appertains to them the highest degree of certainty of which the human mind is capable. It is thought that a valid mathematical theorem is necessarily of such a character as to compel belief in any mind capable of following the steps of the demonstration. Any considerations tending to weaken this belief would be disconcerting and would cause some degree of astonishment. At the risk of this, I must here mention two facts which are of considerable importance as regards an estimation of the cause precise character of mathematical knowledge.

In the first place, it is a fact that frequently, and at various times, differences of opinion have existed among mathematicians, giving rice to controversies as to the validity of whole lines of reasoning and affecting the results of such reasoning; a considerable amount of difference of opinion of this character exists among mathematicians at the present time.

In the second place, the accepted standard of rigour, that is, the standard of what is deemed necessary to constitute a valid demonstration, has undergone change in the course of time. Much of the reasoning which was formerly regarded as satisfactory and irrefutable is now regarded as insufficient to establish the results which it was employed to demonstrate. It has even been shown that results which were once supposed to have been fully established by demonstrations are, in point of fact, affected with error. I propose here to explain in general terms how these phenomena are possible.

In every subject of study, if one probes deep enough, there are found to be points in which that subject comes in contact with general philosophy, and where differences of philosophical view will have a greater or less influence on the attitude of the mind towards the principles of the particular subject. This is not surprising when we reflect that there is but one universe of thought, that no department of knowledge can be absolutely isolated, and that metaphysical and psychological implications are a necessary element in all the activities of the mind. A particular department, such as Mathematics, is compelled to set up a more or less artificial frontier, which marks it off from general philosophy. This frontier consists of a set of regulative ideas in the form of indefinables and axioms, partly ontological assumptions, and partly postulations of a logical character. To go behind these, to attempt to analyse their nature and origin, and to justify their validity, is to go outside the special department and to touch on the domains of the metaphysician and the psychologist. Whether they are regarded as possessing apodictic certainty or as purely hypothetical in character, these ideas represent the data or premisses of the science, and the whole of its edifice is dependent upon them. They serve as the foundation on which all is built, as well as the frontier on the side of philosophy and psychology. A set of data ideally perfect in respect of precision and permanence is unattainable—or at least has not yet been attained; and the adjustment of frontiers is one of the most frequent causes of strife. As a matter of fact, variations of opinion have at various times arisen within the ranks of the mathematicians as to the nature, scope, and proper formulation of the principles which form the foundations of the science, and the views of mathematicians in this regard have always necessarily been largely affected by the conscious or unconscious attitude of particular minds towards questions of general philosophy. It is in this region, I think, that the source is to be found of those remarkable differences of opinion amongst mathematicians which have come into prominence at various times, and have given rise to much controversy as to fundamentals.

II. The following excerpts are drawn from *"Squaring the Circle,"*[43] a monograph Hobson authored shortly before the outbreak of the First World War:

A general survey of the history of thought reveals to us the fact of the existence of various questions that have occupied the almost continuous attention of the thinking part of mankind for long series of centuries. Certain fundamental questions presented themselves to the human mind at the dawn of the history of speculative thought, and have maintained their substantial identity throughout the centuries, although the precise terms in which such questions have been stated have varied from age to age in accordance with the ever varying attitude of mankind towards fundamentals. In general, it may be maintained that, to such questions, even after thousands of years of discussion, no answers have been given that have permanently satisfied the thinking world, or that have been generally accepted as final solutions of the matters concerned. It has been said that those problems that have the longest history are the insoluble ones.

If the contemplation of this kind of relative failure of the efforts of the human mind is calculated to produce a certain sense of depression, it may be a relief to turn to certain problems, albeit in a more restricted domain, that have occupied the minds of men for thousands of years, but which have at last, in the course of the

43 E. W. Hobson, *"Squaring the Circle": A History of the Problem* (Cambridge University Press, 1913; pp. 1-4. Large blocks of text have been broken up into smaller paragraphs for ease of reading.

nineteenth century, received solutions that we have reasons of overwhelming cogency to regard as final. Success, even in a comparatively limited field, is some compensation for failure in a wider field of endeavour. Our legitimate satisfaction at such exceptional success is but slightly qualified by the fact that the answers ultimately reached are in a certain sense of a negative character. We may rest contented with proofs that these problems, in their original somewhat narrow form, are insoluble, provided we attain, as is actually the case in some celebrated instances, to a complete comprehension of the grounds, resting upon a thoroughly established theoretical basis, upon which our final conviction of the insolubility of the problems is founded.

The three celebrated problems of the quadrature of the circle, the trisection of an angle, and the duplication of the cube, although all of them are somewhat special in character, have one great advantage for the purposes of historical study, viz. that their complete history as scientific problems lies, in a completed form, before us. Taking the first of these problems, which will be here our special subject of study, we possess indications of its origin in remote antiquity, we are able to follow the lines on which the treatment of the problem proceeded and changed from age to age in accordance with the progressive development of general Mathematical Science, on which it exercised a noticeable reaction. We are also able to see how the progress of endeavours towards a solution was affected by the intervention of some of the greatest Mathematical thinkers that the world has seen, such men as Archimedes, Huyghens, Euler, and Hermite. Lastly, we know when and how the resources of modern Mathematical Science became sufficiently powerful to make possible that resolution of the problem which, although negative, in that the impossibility of the problem subject to the implied restrictions was proved, is far from being a mere negation, in that the true grounds of the impossibility have been set forth with a finality and completeness which is somewhat rare in the history of Science...

The popularity of the problem among non-Mathematicians may seem to require some explanation. No doubt, the fact of its comparative obviousness explains in part at least its popularity; unlike many Mathematical problems, its nature can in some sense be understood by anyone; although, as we shall presently see, the very terms in which it is usually stated tend to suggest an imperfect apprehension of its precise import. The accumulated celebrity which the problem attained, as one of proverbial difficulty, makes it an irresistible attraction to men with a certain kind of mentality. An exaggerated notion of the gain which would accrue to mankind by a solution of the problem has at various times been a factor in stimulating the efforts of men with more zeal than knowledge. The man of mystical tendencies has been attracted to the problem by a vague idea that its solution would, in some dimly discerned manner, prove a key to a knowledge of the inner connections of things far beyond those with which the problem is immediately connected.

III. The following extended excerpt is drawn from Hobson's Gifford Lectures[44] delivered in 1921 and 1922 in Scotland:

In the time of the Greek Mathematicians, and also during the progressive period of Mathematical thought which commenced in the

44 E. W. Hobson, *The Domain of Natural Science: The Gifford Lectures Delivered in the University of Aberdeen in 1921 and 1922* (Cambridge University Press, 1923), pp 117-122. Large blocks of text have been broken up into smaller paragraphs for ease of reading.

sixteenth century and has lasted to the present time, the conceptions relating to the infinite and infinitesimal, together with the related conception of a limit, have almost continuously occupied the attention of Mathematicians and Philosophers.

Much of the confusion of ideas on these matters which lasted from the time of the origin of the Differential Calculus, created by Newton and Leibniz, was due to a failure to distinguish with sufficient clearness of outline between the conceptual and the perceptual sides of the measurement of magnitudes, and to the uncritical acceptance of notions derived from sensuous intuition as sufficient for the basis of a rigorous conceptual scheme. On the subject of the infinite and the infinitesimal the views expressed have frequently presented a diversity akin to that which has been exhibited in relation to general Philosophy. There have been all shades of believers, sceptics, pragmatists, and finitists.

Although it is a popular belief that Mathematics is of such a character as to leave no room for differences of opinion, it is a fact that, even at the present time, there exist differences of opinion amongst Mathematicians about the foundations of the Science, and more especially about matters in which the notion of the infinite is involved. In the earlier presentations of the Differential and Integral Calculus, frequently called the Infinitesimal Calculus, theories concerning infinitesimal magnitudes played a large part. The notion of an infinitesimal was frequently complicated with ideas about motion, derived uncritically from spatial and temporal intuition. Resulting from the clarification of the conceptions of the foundations of Arithmetic and Mathematical Analysis, resting upon a purely arithmetical basis, which has taken place during the last half century, the conception of infinitesimal numbers has been excluded from ordinary Arithmetical Analysis, as an unnecessary conception in the scheme. Every number regarded, for the purposes of the Calculus, as existent, belongs to the arithmetic continuum, and is therefore finite, if it be not the single number zero. The term *infinitesimal* is no longer used to denote a number, or a magnitude; when it is used at all it is employed to describe a process of change, and even for that purpose it is better avoided, so as not to give rise to misunderstandings.

The fundamental conception of a limit, to which I have already referred in connection with the definition of an irrational number, emerged first amongst the Greeks, in a geometrical form, embodied in the method of exhaustions. It is remarkable that their conception of the nature of a limit attained a standard of rigour, greater than that which obtained amongst modern Mathematicians before our own time. This is exemplified in the proof given by Euclid, in the eleventh book of his Elements, that the circumference of a circle varies as its radius. Archimedes applied the method of limits in a rigorous manner to various problems of what we now call integration, such as the determination of areas and volumes.

The advantages which the methods of Leibniz and of Newton had over that of Archimedes were of a practical kind, as they had a form which made them readily applicable to calculations, whereas the geometrical form in which the Greek method was clothed, together with the absence of a convenient arithmetical scheme, made the application of the Greek method to particular problems decidedly cumbersome. But in the matter of soundness of theory, the method of Archimedes was superior to those of Leibniz and Newton. The construction of the arithmetic continuum was an absolutely necessary requisite for a rigorous theory of limits, and thus for the foundation, on a logical basis, of the Differential and Integral Calculus. In default of such a construction, when the notion of magnitude, as directly given by sensuous intuition, was taken as part of the basis of the doctrine of limits, the frequent endeavours that were made to prove that every convergent sequence of numbers necessarily converges to a limit were doomed to inevitable failure.

That the contemplation of the infinite, in some form, is indispensable to Mathematicians arises from the fact that even comparatively simple problems, such as those of the determination of the length of a curvilinear arc, or of the area enclosed by a curve, can in no case be solved by employing a finite number of the operations of arithmetic, except when the curve consists of segments of straight lines. An approximation to the magnitude of an area can be found by division of the area into a sufficiently large number of rectangles, leaving a small undetermined part of the area out of account.

By increasing the number of rectangles indefinitely in a suitable manner, the measure of the area required is exhibited as the limit of the sequence of numbers determined by the approximations. Thus the magnitude to be determined is only obtainable as the limit of a sequence involving an indefinitely great set of arithmetical operations. The Integral Calculus is concerned with methods of calculating the limits defined in such a manner. It is clear moreover that the very conception of an area or length, as having a definite magnitude, is dependent upon the concept of a limit, and that it is only defined by, and is dependent upon, the existence of a definite limit to a sequence.

The notion of the gradient, or differential coefficient of one variable with respect to another, is one which is essentially dependent upon the conception of a limit, and the gradient only has a precise meaning when a rigorous theory of limits has been previously established. The older idea that a gradient may be regarded as the limiting ratio of vanishing quantities, and which was justly derided by Bishop Berkeley,[45] affords a striking example of the hazy conceptions of fundamental matters with which Mathematicians for a long period of time contented themselves.

The abstract conceptual schemes, of an advanced character, employed for the purpose of representing measurable physical processes, are very frequently expressed in the form of relations between the gradients of various magnitudes with respect to other magnitudes, or to the time-variable. These relations, known as differential equations, are of fundamental importance in connection with such abstract schemes of representation; and the mathematical theory concerned with the determination of the values of variables, under certain conditions, in terms of the time-variable, is consequently, for the purposes of theoretical physics, of great importance. In the theory of differential equations, we have one of the most important examples of the fact that the abstract theories of Pure Mathematics provide the means for utilizing conceptual scientific schemes for the purpose of representing in their quantitative aspect a great variety of physical phenomena.

The fact that Mathematical methods are, in a very large class of cases, unable to deal with objects, or with processes, except by breaking them up into parts, and increasing indefinitely the number of those parts, is a significant example of a limitation imposed upon us by what appears to be a definite characteristic of our modes of

45 George Berkeley, *The Analyst* (1734): "They are neither finite Quantities, nor Quantities infinitely small, nor yet nothing. May we not call them the Ghosts of departed Quantities?"

apprehension. We appear to be unable to grasp some of the relations of a whole, without breaking it up, as it were atomistically, and then proceeding to reconstruct the whole by a synthetic process which is confined to a continual approach to the whole along the path of an endless regress, which by its very nature, is such that the whole is never actually reached within the process, although a scrutiny of the laws of the regress may enable us to obtain a knowledge of the relations of the whole.

The mathematical theory of the numerically infinite, and especially the developments of Georg Cantor and his school in the theory of infinite aggregates of objects, have aroused considerable interest in the ranks of Philosophers. Some of them, such as, for example, Josiah Royce,[46] have suggested that the results of the theory of transfinite numbers may throw light upon questions of General Philosophy, such as the fundamental problem of the One and the Many. I cannot enter here into a discussion of the extension of the theory of the transfinite into the more general region of thought, but I would suggest that extreme caution should be exercised in attempting to extend results of such a theory as that of transfinite aggregates to a domain wider than its original one. The theory has been created for a special purpose, that of dealing with certain aspects of the numerically infinite, and its constructions and results are all dependent upon a set of postulations and definitions which it has been the aim of investigators to make as precise in character as possible. The most careful scrutiny of the meaning to

46 Josiah Royce delivered the Gifford Lectures in 1899. See the Supplementary Essay, "The One, The Many, and the Infinite," in Royce, *The World and the Individual: Gifford Lectures Delivered before the University of Aberdeen*, First Series (The Macmillan Company, 1900).

be attached to the terms employed in any extensions of the theory, such as those to which I have alluded, is of the last importance; otherwise there is a serious danger of falling into grave errors in setting up theories in which vague analogies, involving the surreptitious use of such terms as the infinite, in a sense different from that in which they are employed in the Mathematical theory, take the place which should be occupied by a critically explored foundation.

David Eugene Smith (1860-1944)

The Science Venerable

Voltaire once remarked—"One merit of poetry few will deny; it says more and in fewer words than prose." With equal significance we may say, "One merit of mathematics few will deny; it says more and in fewer words than any other science." The formula $e^{\pi i} = -1$ expresses a world of thought, of truth, of poetry and of religious spirit, for "God eternally geometrizes."

— Creed of David Eugene Smith

Made in United States
Orlando, FL
04 November 2026

90205461R00053